The Enigma of Eternity

Unraveling the Secrets of Time and Space

Steffen Fiedler

All rights reserved. No part of this publication may be reproduced, distributed, or transmitted in any form or by any means, including photocopying, recording, or other electronic or mechanical methods, without the prior written permission of the publisher, except in the case of brief quotations embodied in critical reviews and certain other noncommercial uses permitted by copyright law.

Copyright © Steffen Fiedler, 2023.

Table of contents

Introduction
Chapter 1
Chapter 2
Chapter 3
Chapter 4
Chapter 5
Chapter 6
Chapter 7
Conclusion

Introduction

The Intriguing Nature of Time and Space

Time and space have captivated human curiosity and sparked countless discussions and inquiries throughout history. Their intriguing nature lies in their fundamental role in shaping our perception of reality and the universe we inhabit. Let's explore some of the fascinating aspects of time and space.

1. Relativity: Albert Einstein revolutionized our understanding of space and time with his theory of relativity. According to the theory, space and time are not absolute but rather intertwined in a unified framework called spacetime. It states that the perception of time can vary depending on the relative motion of observers and the strength of gravitational fields. This concept challenges our intuitive notions of a fixed, universal time.

2. Time Dilation: One of the remarkable consequences of relativity is time dilation. It suggests that time can appear to pass at different rates for objects moving at different speeds or experiencing different gravitational forces. For instance, astronauts on the International Space Station experience time dilation, aging slightly slower compared to people on Earth due to their higher orbital speed and weaker gravity.

3. Black Holes: Black holes are extraordinary cosmic objects where gravity is so intense that nothing, not even light, can escape their gravitational pull. They form when massive stars collapse under their own gravity. Near a black hole's event horizon (the boundary beyond which nothing can escape), time is dramatically affected. Observers from afar would perceive time for someone falling into a black hole as slowing down and eventually freezing at the event horizon.

4. Wormholes and Time Travel: The concept of wormholes, hypothetical tunnels in spacetime, adds another layer of intrigue to the nature of time and space. While purely speculative, wormholes present a theoretical possibility for shortcuts between distant parts of the universe or even different points in time. If traversable wormholes existed, they could potentially allow for time travel or at least the transmission of information across time.

5. Expansion of the Universe: The Big Bang theory suggests that the universe originated from an extremely dense and hot state approximately 13.8 billion years ago. Since then, the universe has been expanding. This expansion means that the fabric of space itself is stretching, carrying galaxies away from each other. It raises questions about the ultimate fate of the universe and the concept of an "edge" or boundary in space.

6. Multiverse and Parallel Universes: The possibility of a multiverse, a collection of parallel universes with different physical laws or conditions, has gained attention in scientific and philosophical discussions. Some theories propose that our universe is just one among many, each with its own set of physical constants and laws of nature. Exploring the existence of parallel universes would have profound implications for our understanding of time and space.

7. Quantum Entanglement: Quantum mechanics, the branch of physics that describes the behavior of particles on the smallest scales, presents another puzzling aspect of space and time. Quantum entanglement refers to the phenomenon where particles become linked together in such a way that their states are intimately connected, regardless of the distance between them. This instantaneous correlation challenges our conventional understanding of space and the speed at which information can be transmitted.

The nature of time and space continues to be a source of fascination, prompting ongoing scientific research, philosophical debates, and inspiring works of art. While we have made significant strides in unraveling their mysteries, many questions remain unanswered, leaving ample room for further exploration and discovery.

- Setting the stage for the exploration of timeless questions

The exploration of timeless questions about time and space has captivated human minds throughout history. As we delve into the depths of these enigmatic concepts, we find ourselves contemplating profound inquiries that transcend the boundaries of our everyday experiences. Here are some key aspects that set the stage for such exploration:

1. Philosophical Reflection: Philosophers have pondered the nature of time and space since ancient times. Questions about the nature of existence, the nature of reality, and the relationship between the physical world and our perception of it have driven philosophical discourse. The exploration of timeless questions encourages deep reflection on the fundamental nature of our reality and our place within it.

2. Scientific Inquiry: Science provides a systematic approach to understanding the mysteries of time and space. Through observation, experimentation, and theoretical frameworks, scientists have sought to unravel the laws governing the universe. Scientific discoveries, such as relativity and quantum mechanics, have revolutionized our understanding of time and space, prompting further investigation and pushing the boundaries of our knowledge.

3. Technological Advancements: Technological advancements have played a crucial role in our exploration of time and space. From the invention of telescopes that allowed us to peer into the vastness of the cosmos to particle accelerators that unlock the secrets of the microscopic world, our tools have expanded our capacity to observe, measure, and probe the nature of reality.

4. Interdisciplinary Collaboration: The exploration of timeless questions often transcends disciplinary boundaries. Collaborations between scientists, philosophers, artists, and thinkers from various fields can foster new perspectives and insights. The integration of diverse approaches and disciplines enriches our exploration of time and space, enabling us to tackle complex problems from different angles.

5. Cultural and Artistic Expressions: Time and space have long inspired artistic expression. Through literature, music, visual arts, and other forms of creativity, artists attempt to capture the ineffable aspects of existence, time's passage, and the vastness of space. Artistic representations can provoke contemplation, evoke emotions, and offer unique perspectives on timeless questions.

6. Human Curiosity and Wonder: At the heart of the exploration of timeless questions lies human curiosity and wonder. Our innate drive to understand the world around us and our place within it fuels the quest for knowledge. The mysteries of time and space evoke awe and fascination, pushing us to seek answers, challenge assumptions, and continuously expand our understanding.

In summary, the exploration of timeless questions about time and space is propelled by philosophical reflection, scientific inquiry, technological advancements, interdisciplinary collaboration, cultural expressions, and the innate human curiosity to unravel the mysteries of existence. As we embark on this exploration, we enter a realm where the boundaries of knowledge are pushed further, and new insights and discoveries await.

- The relevance of time and space in our everyday lives

Time and space play significant roles in shaping our everyday lives in various ways. Here are some key aspects that highlight their relevance:

1. Organization and Scheduling: Time provides the framework for organizing and scheduling our daily activities. It helps us plan our routines, set deadlines, and allocate resources efficiently. Concepts like hours, minutes, and seconds structure our lives, allowing us to coordinate with others, meet obligations, and accomplish tasks in a timely manner.

2. Personal Productivity: Understanding and managing time effectively is crucial for personal productivity. Being aware of how we spend our time, setting priorities, and practicing time management techniques can enhance our efficiency and achievement of goals. Time management skills enable us to balance work, leisure, relationships, and personal development, leading to a more fulfilling and well-rounded life.

3. Relationships and Communication: Time influences our interactions and relationships with others. It allows us to coordinate meetings, schedule gatherings, and maintain social connections. Punctuality and respecting others' time are essential for effective communication and building trust in personal and professional relationships.

4. Spatial Orientation and Navigation: Space is integral to our ability to navigate and orient ourselves in the physical world. We rely on spatial awareness to find our way, whether it's through streets, buildings, or natural environments. Understanding spatial relationships helps us create mental maps, use GPS systems, and explore new places with confidence.

5. Travel and Transportation: Time and space are crucial considerations when it comes to travel and transportation. The ability to estimate travel time helps us plan journeys, make arrangements, and manage logistics. Understanding spatial distances and routes enables us to navigate transportation systems, choose the most efficient paths, and reach our destinations efficiently.

6. Cultural and Social Context: Time and space have cultural and social dimensions that shape our behaviors and interactions. Different cultures may have varying perceptions of time, such as prioritizing punctuality or adopting more relaxed attitudes. Spatial arrangements, such as the design of public spaces or private homes, can influence social dynamics, privacy, and community engagement.

7. Personal Reflection and Well-being: Time and space provide opportunities for personal reflection, introspection, and well-being. Allocating time for relaxation, leisure activities, hobbies, and mindfulness practices allows us to recharge, reflect on our experiences, and nurture our mental and emotional well-being. Creating physical spaces that support relaxation and personal growth contributes to our overall quality of life.

8. Historical and Future Perspectives: Time and space connect us to the broader context of history and the future. Studying history helps us understand our past, learn from it, and appreciate the continuity of human experiences across time. Considering the future in terms of long-term planning and aspirations enables us to make informed decisions and shape our individual and collective paths.

In summary, time and space are intertwined in our everyday lives, influencing how we organize our activities, interact with others, navigate our surroundings, and reflect on our experiences. Recognizing their relevance allows us to optimize our use of time, appreciate the spaces we inhabit, and cultivate a sense of balance, purpose, and well-being.

Chapter 1

The Evolution of Time: From Ancient Perspectives to Modern Theories

The concept of time has evolved significantly over centuries, as various civilizations and thinkers have grappled with its nature and sought to understand its essence. Let's explore the evolution of time from ancient perspectives to modern theories:

Ancient Perspectives:
1. Cyclical Time: Many ancient cultures, such as the ancient Egyptians and Mayans, viewed time as cyclical, with events recurring in predictable patterns. This perspective was often tied to natural phenomena, such as the rising and setting of the sun or the changing seasons.

2. Mythological and Religious Time: Time was often intertwined with mythological or religious beliefs. Ancient civilizations, like the Greeks and Romans, had gods and goddesses associated with time, such as Chronos and Saturn. Time was seen as part of the divine order and a reflection of cosmic cycles.

3. Linear Time: In contrast to cyclical time, some ancient cultures, such as the Hebrews, introduced the notion of linear time, where events progressed in a linear fashion, leading to a final culmination or purpose. This idea laid the foundation for the concept of historical progression.

Medieval and Renaissance Views:
1. Divine Time: During the Middle Ages, time was often seen as a reflection of God's will and divine plan. The idea of an eternal and unchanging God contrasted with the finite and constantly changing nature of time in the earthly realm.

2. Cosmic Time: The Renaissance brought a resurgence of interest in science and a shift towards understanding time in relation to the physical world. Thinkers like Galileo Galilei and

Johannes Kepler explored the idea of time as a component of the cosmic order, influenced by celestial motions.

Enlightenment and Industrial Revolution:
1. Absolute Time: With the advent of modern science and the Enlightenment, philosophers like Isaac Newton proposed the concept of absolute time, independent of any observer or external influences. Newton's idea provided a framework for understanding the laws of motion and laid the groundwork for classical physics.

2. Objective Time: Time became associated with measurement and quantification, enabling greater precision and standardization. The development of clocks and the establishment of time zones during the Industrial Revolution further shaped the perception of time as an objective and uniform entity.

Modern Theories:
1. Relativity: In the early 20th century, Albert Einstein's theory of relativity revolutionized our

understanding of time. The theory introduced the concept of spacetime, where time and space are interconnected and influenced by the relative motion of observers and gravitational fields. It challenged the notion of a universal, absolute time.

2. *Subjective Time:* Modern psychology and neuroscience have explored subjective experiences of time. The perception of time can vary based on attention, emotions, and cognitive processes. Research in this field has shed light on the complex relationship between subjective experience and objective measurements of time.

3. *Quantum Time:* In the realm of quantum mechanics, time assumes a different character. Quantum theory suggests that the flow of time is not fundamental but emerges from the entanglement and interactions of quantum systems. This area of research is still evolving and aims to reconcile quantum mechanics with the nature of time.

The evolution of time from ancient perspectives to modern theories reflects humanity's continuous quest to comprehend and explain this enigmatic concept. As scientific, philosophical, and cultural understandings evolve, our perception of time continues to deepen, posing new questions and expanding the frontiers of knowledge.

- Ancient concepts of time and their influence

Ancient civilizations developed various concepts of time that profoundly influenced their cultures, belief systems, and societal structures. Here are some significant ancient concepts of time and their influence:

1. Cyclical Time: Ancient cultures often viewed time as cyclical, with events recurring in predictable patterns. This concept was closely tied to natural phenomena such as the rising and setting of the sun, the phases of the moon, and the changing seasons. It instilled a sense of continuity, balance, and harmony in the cosmic order. The idea of cyclical time influenced religious rituals, agricultural practices, and the development of calendars.

2. Mythological Time: Many ancient civilizations associated time with their mythologies and religious beliefs. Gods and goddesses were often associated with time, such as the Greek god Chronos or the Roman god Saturn. Time was seen as part of the divine realm and a reflection of cosmic cycles. This perspective shaped religious rituals, festivals, and the perception of human existence as part of a larger cosmic narrative.

3. Linear Time: In contrast to cyclical time, some ancient cultures introduced the notion of linear time, where events progressed in a linear fashion, leading to a final culmination or purpose. This concept emerged with the development of historical narratives and the recording of significant events. The idea of linear time influenced the formation of historical accounts, the progression of societies, and the development of ideas like progress, evolution, and teleology.

4. Sacred Time: Time was often regarded as sacred in many ancient cultures. It had a spiritual and mystical dimension, connecting the earthly realm with the divine. Specific moments, such as solstices or equinoxes, were considered sacred and marked with religious ceremonies. This perception of time fostered a sense of reverence, ritual observance, and the belief in the transcendental nature of temporal experiences.

5. Astrological Time: Ancient civilizations, such as the Babylonians and the Mayans, associated time with celestial movements and the positions of stars and planets. They developed intricate systems of astrology, where the alignment of celestial bodies was believed to influence human affairs and destiny. This concept influenced the development of astronomy, astrology, and the belief in a cosmic order governing human lives.

6. Cosmic Time: The concept of cosmic time emerged in various ancient cultures, especially during the Renaissance period. Thinkers like Galileo Galilei and Johannes Kepler explored the idea of time as a component of the cosmic order, influenced by celestial motions. This perspective laid the foundation for the scientific understanding of time and its relationship to the physical world.

Ancient concepts of time not only shaped religious and philosophical beliefs but also influenced practical aspects of daily life, social structures, and cultural practices. They provided frameworks for understanding the passage of time, organizing societies, and establishing calendars. While our contemporary understanding of time has evolved, ancient concepts continue to resonate in various cultural, religious, and spiritual contexts, reflecting the enduring influence of these ancient ideas.

- The emergence of scientific theories and paradigm shifts

The emergence of scientific theories and paradigm shifts represents significant milestones in the progress of scientific knowledge. These shifts occur when new theories challenge and replace existing scientific frameworks, leading to transformative changes in how we understand the natural world. Here's an overview of the process and impact of paradigm shifts:

1. Normal Science: Scientific progress typically occurs within a dominant framework called a paradigm. Normal science refers to the period when scientists work within an established paradigm, conducting experiments, making observations, and refining existing theories. During this phase, scientists aim to solve puzzles and anomalies within the framework of the prevailing paradigm.

2. Anomalies and Crisis: Over time, anomalies—observations or experimental results that do not fit within the current paradigm—may accumulate. These anomalies challenge the existing theories and raise doubts about the paradigm's ability to explain all aspects of the natural world. The accumulation of anomalies can lead to a crisis, where the prevailing paradigm becomes increasingly questioned.

3. Paradigm Shift: A paradigm shift occurs when a new theory or framework emerges that can account for the anomalies and offers a more comprehensive explanation of the phenomena under investigation. This shift marks a transformative change in scientific understanding, often associated with a change in underlying assumptions, methodologies, and conceptual frameworks. Paradigm shifts bring about new perspectives and open up new avenues for scientific inquiry.

4. Scientific Revolution: A paradigm shift is often followed by a period of scientific revolution. During this phase, scientists reexamine existing knowledge, revise theories, and conduct new experiments to explore the implications of the new paradigm. Scientific revolutions can be disruptive, as they challenge established theories, methodologies, and scientific communities' norms. They can lead to a reevaluation of accepted knowledge and spark debates among scientists.

5. Impacts and Progress: Paradigm shifts and scientific revolutions have far-reaching impacts. They can result in breakthroughs, leading to new discoveries, technological advancements, and a deeper understanding of the natural world. Paradigm shifts also shape scientific disciplines, influencing research directions, methodologies, and the education of future scientists. They drive the growth of scientific knowledge and promote scientific progress.

Notable examples of paradigm shifts include the shift from the geocentric model to the heliocentric model of the solar system (led by Copernicus and Galileo) and the shift from Newtonian physics to Einstein's theory of relativity. These shifts fundamentally transformed our understanding of the universe and had profound effects on subsequent scientific developments.

In summary, paradigm shifts mark transformative moments in the progression of scientific knowledge. They arise when new theories challenge prevailing paradigms, leading to scientific revolutions and shifts in our understanding of the natural world. These shifts drive scientific progress, reshape disciplines, and open up new avenues for exploration and discovery.

Chapter 2

The Physics of Time: Relativity, Quantum Mechanics, and Beyond
Einstein's theory of relativity and its implications for time

Einstein's theory of relativity, particularly his theory of special relativity and the subsequent theory of general relativity, revolutionized our understanding of time and its relationship to space and gravity. Here's an overview of Einstein's theory of relativity and its implications for time:

1. Special Relativity: Einstein's theory of special relativity, proposed in 1905, introduced a new understanding of space and time. It postulated that the laws of physics are the same in all inertial reference frames, and the speed of light in a vacuum is constant for all observers. Key implications of special relativity include:

Time Dilation: Special relativity predicts that time can dilate, or "slow down," depending on an object's relative motion. As an object approaches the speed of light, time appears to pass more slowly for that object compared to a stationary observer. This phenomenon has been experimentally verified and has practical implications, such as in the operation of global positioning systems (GPS).

***<u>Relativistic Length</u* <u>Contraction:</u>** According to special relativity, objects in motion appear to contract in the direction of their motion from the perspective of a stationary observer. This length contraction effect is a consequence of the spacetime geometry predicted by the theory.

2. General Relativity: Einstein's theory of general relativity, published in 1915, expanded upon special relativity by incorporating the effects of gravity. General relativity postulates that gravity arises due to the curvature of spacetime caused by massive objects. Key implications of general relativity include:

Time Dilation in Gravitational Fields:
General relativity predicts that the strength of the gravitational field affects the passage of time. Clocks in stronger gravitational fields (e.g., closer to massive objects) appear to run more slowly compared to clocks in weaker gravitational fields. This effect has been confirmed through experimental observations, such as with atomic clocks on Earth and in space.

Time and Space as a Continuum: General relativity describes the fabric of spacetime as a four-dimensional continuum, where space and time are intertwined. Gravity arises from the curvature of this continuum caused by matter and energy. The theory provides a unified framework for understanding gravity and its influence on the geometry of spacetime.

Beyond Einstein's theories of relativity, the nature of time becomes even more intriguing in the realm of quantum mechanics. While a complete theory of quantum gravity that unifies general relativity and quantum mechanics has yet to be realized, researchers are actively exploring the intersection of these two fundamental theories.

Quantum theories of time suggest that time may not be fundamental but rather emerge from quantum processes or entanglement. Researchers are investigating the nature of time in the context of quantum cosmology, black hole physics, and the study of the early universe.

In summary, Einstein's theory of relativity, encompassing special and general relativity, fundamentally reshaped our understanding of time. It introduced concepts like time dilation and the curvature of spacetime, which have been experimentally verified and are crucial for modern physics. Exploring the connection between relativity and quantum mechanics continues to be an active area of research, with the aim of unraveling the deeper nature of time and its role in the fundamental fabric of the universe.

- Quantum mechanics and its strange effects on our understanding of time

Quantum mechanics, the branch of physics that deals with the behavior of particles at the smallest scales, introduces several intriguing and non-intuitive effects that challenge our conventional understanding of time. Here are some of the strange effects of quantum mechanics on our understanding of time:

1. Superposition: In quantum mechanics, particles can exist in a state of superposition, where they simultaneously occupy multiple possible states. This concept raises questions about the nature of time because it suggests that particles can exist in a superposition of states across different points in time. This challenges the idea of a linear progression of time and suggests a more complex relationship between particles and temporal states.

2. Quantum Entanglement: Quantum entanglement occurs when two or more particles become correlated in such a way that the state of one particle cannot be described independently of the others. This entanglement can persist even when the particles are physically separated. The phenomenon of entanglement raises questions about the flow of time because the correlated particles seem to share a non-local connection that transcends conventional notions of distance and temporal ordering.

3. Time Symmetry: Quantum mechanics suggests that the fundamental laws of physics are time-symmetric, meaning that they remain unchanged if the direction of time is reversed. While macroscopic systems exhibit time asymmetry (e.g., the irreversible nature of processes like aging or the diffusion of gas), at the quantum level, processes can be both forwards and backwards in time. This implies that the concept of cause and effect becomes more nuanced, challenging our intuitive understanding of time as a unidirectional arrow.

4. Time Uncertainty: According to Heisenberg's uncertainty principle, there is a fundamental limit to the precision with which certain pairs of physical properties, such as position and momentum, can be simultaneously known. This principle extends to time and energy, implying that there is an inherent uncertainty in measuring both quantities precisely. This time-energy uncertainty challenges the notion of time as a precisely measurable and well-defined parameter.

5. Quantum Tunneling: Quantum tunneling is a phenomenon where particles can pass through energy barriers that, according to classical physics, they should not be able to overcome. This effect suggests that particles can seemingly "teleport" through barriers without requiring the time it would take to traverse them conventionally. Quantum tunneling raises questions about the continuity and causality of time, as particles can effectively bypass physical barriers without adhering to traditional temporal constraints.

It is important to note that these quantum effects on time are still subjects of ongoing research and investigation. While they challenge our intuitive understanding of time, they also highlight the rich and complex nature of the quantum realm. Understanding the interplay between quantum mechanics and the nature of time remains an active area of exploration, with the aim of deepening our understanding of the fundamental fabric of the universe.

Chapter 3

The Fabric of Spacetime: Unraveling the Mysteries

The concept of the fabric of spacetime lies at the heart of our current understanding of the universe. It combines the notions of space and time into a unified framework, describing how they are interconnected and influenced by matter and energy. Let's delve into the mysteries surrounding the fabric of spacetime:

1. Spacetime as a Continuum: According to Einstein's theory of general relativity, spacetime is not separate entities of space and time but rather a four-dimensional continuum. It treats space and time as inseparable, forming a unified fabric in which all events occur. This fabric is not fixed but can be curved, warped, and influenced by the presence of mass and energy.

2. Curvature and Gravity: In general relativity, the curvature of spacetime is intimately linked to the distribution of mass and energy. Massive objects, such as planets, stars, and galaxies, curve the surrounding spacetime, creating what we perceive as gravitational forces. The curvature of spacetime determines the paths followed by objects moving in its vicinity, including the motion of light.

3. The Geometry of Spacetime: The curvature of spacetime is described by its geometry. In regions with high mass and energy density, spacetime curvature becomes significant, resulting in strong gravitational effects. This understanding of spacetime's geometry allows us to explain phenomena like the bending of starlight around massive objects (gravitational lensing) and the motion of planets in their orbits.

4. Black Holes: One of the most intriguing consequences of the fabric of spacetime is the existence of black holes. Black holes form when massive stars collapse under their own gravity, leading to an extremely curved spacetime region called a singularity. At the singularity, the fabric of spacetime becomes infinitely curved, and the gravitational pull is so strong that nothing, not even light, can escape its grasp.

5. Wormholes and Time Travel: The fabric of spacetime also allows for theoretical possibilities like wormholes, which are shortcuts or tunnels through spacetime. Wormholes, if they exist, could connect distant regions of the universe or even different points in time. Theoretical discussions on wormholes raise questions about the potential for time travel and the fundamental nature of causality within the fabric of spacetime.

6. Quantum Gravity: Understanding the fabric of spacetime at the quantum level remains an ongoing challenge in physics. The quest for a theory of quantum gravity aims to reconcile the principles of quantum mechanics with the curved geometry of spacetime described by general relativity. Such a theory would provide a more complete understanding of the fabric of spacetime on the smallest scales and during the early moments of the universe.

Exploring the mysteries of the fabric of spacetime continues to drive scientific inquiry. Through ongoing research, observational studies, and theoretical investigations, scientists strive to deepen our understanding of this fundamental structure of the universe, unravel its complexities, and shed light on the nature of gravity, the origin of the cosmos, and the fundamental laws that govern our existence.

- Exploring the concept of spacetime and its interconnectedness

The concept of spacetime combines the notions of space and time into a unified framework, highlighting their interconnectedness. In this framework, events in the universe are described by their coordinates in both space and time.

Here, we'll further explore the concept of spacetime and its interconnected nature:

1. Minkowski Spacetime: The mathematical formulation of spacetime was first introduced by the mathematician Hermann Minkowski. He combined the three dimensions of space (length, width, and height) with the dimension of time to create a four-dimensional structure known as Minkowski spacetime. In this framework, each event in the universe is characterized by four coordinates: three for space and one for time.

2. Inseparability of Space and Time: In the framework of spacetime, space and time are inseparable. They are not separate entities but different aspects of the same underlying fabric. Just as an object's position in space can change, so can its position in time. Spacetime unifies the concepts of spatial dimensions and the dimension of time, providing a holistic framework for describing the dynamics of the universe.

3. Relativistic Effects: The interconnectedness of space and time becomes evident through relativistic effects, as predicted by Einstein's theory of relativity. Special relativity demonstrates that space and time intervals are not absolute but depend on the relative motion of observers. For example, time can appear to pass more slowly for an object in motion compared to a stationary observer. This time dilation effect is a consequence of the interplay between space and time in spacetime.

4. Spacetime Curvature: According to general relativity, the presence of mass and energy curves the fabric of spacetime, resulting in what we perceive as gravity. The curvature of spacetime is influenced by the distribution of matter and determines the motion of objects within it. Massive objects create a "dent" in spacetime, causing other objects to follow curved paths around them. This curvature illustrates the interconnected nature of space and time and how they are influenced by the presence of mass and energy.

5. Light Cones: In the framework of spacetime, light cones play a crucial role in understanding causality and the flow of information. A light cone is a visual representation of the paths that light (or any other signal) can take in spacetime. It consists of two components: the future light cone, representing events that can be influenced by a given event, and the past light cone, representing events that can influence a given event. The shape of the light cones emphasizes the interconnectedness of events in spacetime and the limitations imposed by the speed of light as a maximum signal velocity.

6. Spacetime as a Dynamic Entity: Spacetime is not a static backdrop but a dynamic entity that can be influenced by the distribution of matter and energy. Changes in the distribution of mass and energy can lead to the curvature of spacetime, affecting the motion of objects and the behavior of physical phenomena. The concept of spacetime as a dynamic entity highlights its interconnected nature with the evolving structure of the universe.

Exploring the interconnectedness of space and time in the framework of spacetime provides a deeper understanding of the fundamental fabric of the universe. It allows us to unravel the intricate relationships between events, the influence of gravity, and the dynamics of particles and fields. The concept of spacetime has revolutionized our understanding of the universe and continues to be a central pillar of modern physics.

- Gravity, curvature, and the warping of spacetime

In Einstein's theory of general relativity, the presence of mass and energy causes spacetime to curve, giving rise to what we perceive as gravity. This concept of gravity as the curvature of spacetime provides a deeper understanding of how massive objects influence the fabric of the universe. Let's explore gravity, curvature, and the warping of spacetime:

1. Gravity as Curvature: According to general relativity, gravity is not a force acting at a distance but a result of the curvature of spacetime caused by mass and energy. Massive objects, such as stars or planets, curve the surrounding spacetime, creating a gravitational field. Other objects moving in this curved spacetime follow curved paths, giving the appearance of being attracted to the massive object. In this framework, objects move along the shortest paths (geodesics) in the curved spacetime geometry.

2. Warping of Spacetime: The presence of mass and energy warps the fabric of spacetime, altering the geometry of the surrounding region. Think of spacetime as a flexible, rubber-like sheet that curves under the influence of mass and energy. The more massive an object, the greater the curvature it induces in the surrounding spacetime. This warping of spacetime represents the gravitational field associated with the object.

3. Geodesics and Curved Paths: In the curved spacetime near a massive object, the paths of objects, including light, follow curved trajectories. These paths are known as geodesics. For example, the motion of planets around the Sun is a result of following the curved paths dictated by the warped spacetime geometry caused by the Sun's mass.

4. Gravitational Time Dilation: The warping of spacetime due to gravity also influences the flow of time. Clocks in regions of stronger gravitational fields, where spacetime is more curved, tick more slowly compared to clocks in weaker gravitational fields. This effect, known as gravitational time dilation, has been confirmed by experiments, including those involving precise atomic clocks on Earth and in space. It demonstrates that gravity affects not only the geometry of spacetime but also the passage of time itself.

5. Black Holes: The most extreme manifestation of spacetime curvature occurs in black holes. Black holes are regions where the curvature of spacetime becomes so intense that nothing, not even light, can escape their gravitational pull. At the center of a black hole lies a singularity, where spacetime curvature becomes infinitely strong. The formation and behavior of black holes are a consequence of the warping of spacetime caused by massive objects collapsing under their own gravity.

Understanding the interplay between gravity, curvature, and the warping of spacetime is a cornerstone of general relativity. This framework provides a comprehensive description of gravity as a geometrical phenomenon, explaining the motion of celestial objects, the bending of light, and the effects of time dilation. The concept of spacetime curvature has been confirmed by numerous observational tests, further solidifying our understanding of gravity as the warping of the fabric of the universe.

Chapter 4

Time Travel: Fact or Fiction?

The concept of time travel has captured the imagination of people for centuries, fueling numerous works of science fiction. However, when it comes to the question of whether time travel is possible in reality, the scientific consensus is that we do not yet have evidence or a theoretical framework that supports the existence of time travel as commonly depicted in fiction. Let's explore this further:

1. Time Travel in Fiction: Time travel has been a popular theme in literature, movies, and television, often presenting different scenarios and mechanisms for traversing time. These fictional portrayals range from using advanced technology and machines to supernatural or fantastical means of time manipulation. While these stories can be entertaining and thought-provoking, they are works of fiction and not based on current scientific understanding.

2. Theoretical Possibilities: While we cannot definitively rule out the possibility of time travel, theoretical physics provides some potential avenues for exploring the concept. One such concept is the idea of closed timelike curves (CTCs) within certain solutions of Einstein's equations of general relativity. CTCs would theoretically allow for closed loops in spacetime that could potentially enable time travel. However, the existence and practicality of CTCs are still highly speculative, and their realization would require exotic forms of matter and energy that have not been observed or understood.

3. Paradoxes and Causality: One of the major challenges with the concept of time travel is the potential for paradoxes and violations of causality. For example, the famous "grandfather paradox" arises when a time traveler goes back in time and prevents their own existence by altering events in the past. Resolving such paradoxes within a consistent framework of cause and effect is a significant challenge.

4. Temporal Consistency Protection: Some physicists propose that the laws of physics could inherently prevent situations that lead to paradoxes. The idea of "temporal consistency protection" suggests that any attempt to change the past would be counteracted by natural forces, ensuring that the timeline remains consistent. This idea is based on various theories, including the Novikov self-consistency principle, which asserts that the laws of physics conspire to prevent changes to the past that would lead to paradoxes.

5. Time Travel and Quantum Mechanics: Quantum mechanics, the theory that describes the behavior of particles at the smallest scales, introduces some intriguing possibilities regarding time. Certain interpretations of quantum mechanics, such as the Many-Worlds

Interpretation, propose the existence of multiple parallel universes, potentially allowing for different "branches" of reality and the notion of "time travel" between these branches. However, these ideas remain highly speculative and are still the subject of scientific investigation and debate.

In summary, while time travel remains an exciting concept in popular culture, the scientific consensus is that we do not currently have evidence or a robust theoretical framework to support the existence of time travel as commonly depicted in fiction. Exploring the possibilities and limitations of time travel is an active area of scientific research, and future advancements in our understanding of physics may shed more light on this intriguing topic.

- Examining the theories and possibilities of time travel

Examining the theories and possibilities of time travel involves delving into various scientific concepts and theoretical frameworks. While time travel as commonly portrayed in fiction remains speculative, let's explore some of the theories and possibilities that have been discussed within the realm of scientific inquiry:

1. Special and General Relativity: Einstein's theories of special relativity and general relativity provide the foundation for understanding the relationship between space, time, and motion. Special relativity introduces the concept of time dilation, where time can appear to pass differently for observers in relative motion. General relativity describes how gravity curves spacetime, potentially allowing for scenarios like time dilation near massive objects or the existence of closed timelike curves under certain conditions.

2. Wormholes: Wormholes are hypothetical structures that connect different regions of spacetime, potentially allowing for shortcuts or tunnels through space and time. If traversable wormholes exist, they could provide a means of traveling to different points in time. However, the feasibility and stability of wormholes remain uncertain, as they would require exotic forms of matter with properties not yet observed.

3. Time Dilation and Time Machines: Time dilation, as predicted by special relativity, suggests that time can pass at different rates depending on the relative motion or the strength of gravitational fields. In theory, traveling close to the speed of light or experiencing extreme gravitational fields could result in time dilation effects, where an observer could experience time differently from someone in a different frame of reference. However, harnessing these effects to create a practical time machine remains purely speculative and faces significant technological and theoretical challenges.

4. Quantum Mechanics and Time: Quantum mechanics, the branch of physics that governs the behavior of particles at the microscopic scale, introduces intriguing possibilities regarding time. Some interpretations, such as the "many-worlds" interpretation, propose the existence of multiple parallel universes or branching realities. The idea of quantum time travel suggests that the manipulation of quantum states or interactions across different branches could potentially lead to time-like effects. However, these ideas are highly speculative and subject to ongoing research and debate.

5. Grandfather Paradox and Temporal Consistency: The famous grandfather paradox highlights a potential issue with time travel, as it raises questions of causality and paradoxes. Resolving such paradoxes while maintaining a consistent framework of cause and effect is a significant challenge. Some theories, such as the Novikov self-consistency principle, propose that the laws of physics would prevent changes to the past that could lead to paradoxes, ensuring that the timeline remains consistent.

It's important to note that the theories and possibilities mentioned above are largely speculative and remain topics of ongoing scientific investigation and debate. Time travel, if it were to become a reality, would likely require profound advancements in our understanding of physics and the discovery of new principles and phenomena.

As scientists continue to explore the frontiers of knowledge, our understanding of time, space, and the nature of the universe may evolve, providing new insights into the feasibility and limitations of time travel. However, for now, time travel remains primarily within the realm of imagination and fiction, awaiting further scientific breakthroughs to determine its true potential.

- Paradoxes, causality, and the philosophical implications

The concept of time travel raises intriguing questions about paradoxes, causality, and the philosophical implications associated with altering the past or the potential for self-inconsistencies. Let's explore these aspects further:

1. Grandfather Paradox: The grandfather paradox is a well-known example that arises in discussions about time travel. It involves a hypothetical scenario where a person travels back in time and prevents their own birth by killing their grandfather before their parent is conceived. This paradox highlights the contradiction that arises when an event in the past affects its own causation, leading to a logical inconsistency. Resolving such paradoxes is a significant challenge and remains a subject of philosophical and scientific inquiry.

2. Causality and the Arrow of Time: Time travel often challenges our intuitive understanding of causality, which is the idea that cause precedes effect. In a scenario where the past is altered, the causal sequence of events becomes disrupted, leading to potential inconsistencies. The arrow of time, which denotes the directionality of cause and effect, is a fundamental concept in physics, and time travel scenarios can challenge this fundamental arrow.

3. Novikov Self-Consistency Principle: The Novikov self-consistency principle is a proposed solution to address paradoxes in time travel scenarios. It suggests that the laws of physics would conspire to prevent changes to the past that could result in self-inconsistent or paradoxical situations. According to this principle, any action taken in the past would already be a part of the timeline, ensuring that the outcome is consistent with the events that led to the time travel in the first place. This principle aims to maintain a consistent framework of cause and effect.

4. Free Will and Determinism: Time travel scenarios raise questions about free will and determinism. If time travel were possible and the past could be altered, it challenges the notion of a predetermined future, suggesting that actions taken in the present could change the course of events. It raises philosophical debates about the nature of free will and whether events are predetermined or subject to change.

5. Ontological and Epistemological Considerations: Time travel also poses philosophical questions about the nature of reality and knowledge. It raises issues related to ontological paradoxes, where information or objects exist without a clear origin or cause, as well as epistemological questions regarding the limits of our knowledge and our ability to comprehend or predict the consequences of time travel.

The paradoxes, causality, and philosophical implications associated with time travel provide fertile ground for contemplation and speculation. While science and philosophy continue to explore these concepts, it is important to note that our current understanding of the laws of physics and the nature of time does not definitively support the existence of time travel or provide clear resolutions to the potential paradoxes it presents.

As our understanding of time, space, and the fundamental principles of the universe evolves, these profound questions may find new insights and perspectives. Ultimately, the exploration of time travel serves as a thought experiment that challenges our understanding of the nature of reality, causality, and our place within the fabric of the universe.

Chapter 5

Parallel Universes: Multiple Realities in the Multiverse

The concept of parallel universes, often discussed within the framework of the multiverse theory, proposes the existence of multiple realities beyond our own. While the idea of parallel universes remains speculative, it has gained attention in both scientific and philosophical circles. Let's explore the concept further:

1. Multiverse Theory: The multiverse theory suggests that our universe is just one among many universes that exist in a larger, interconnected structure called the multiverse. Each universe within the multiverse could have different physical laws, constants, or initial conditions, leading to a diverse array of realities. These universes might coexist alongside ours, potentially separated by vast distances or existing in different dimensions or branes.

2. Many-Worlds Interpretation: The Many-Worlds Interpretation (MWI) is a specific interpretation of quantum mechanics that proposes the existence of parallel universes. According to this interpretation, every quantum measurement or event leads to the branching of reality, resulting in the creation of multiple universes. Each universe represents a different outcome or possibility, with all potentialities being realized across the multiverse. MWI suggests that every quantum possibility is actualized, albeit in separate parallel universes.

3. Inflationary Cosmology: The concept of inflation, a period of rapid expansion in the early universe, is a cornerstone of modern cosmology. Inflationary cosmology predicts that during the exponential expansion, new regions of space can emerge and form separate bubble universes. These bubble universes could exist as distinct entities within the larger multiverse, each with its own set of physical properties and initial conditions.

4. String Theory and Brane Worlds: String theory, a theoretical framework aiming to unify all fundamental forces and particles, suggests the existence of extra dimensions beyond the three spatial dimensions we experience. In some formulations of string theory, known as brane worlds, it is proposed that our universe is a three-dimensional "brane" embedded within a higher-dimensional space. Other branes could exist alongside ours, representing different universes or realities within the multiverse.

5. Observational and Experimental Evidence: Currently, there is no direct observational or experimental evidence supporting the existence of parallel universes. The nature of the multiverse makes it difficult to test empirically, as it would require observations or interactions with other universes. However, various cosmological and theoretical considerations have motivated the exploration of the multiverse concept as a possible solution to certain puzzles and questions within physics.

It is important to note that the multiverse theory, including the existence of parallel universes, is a subject of ongoing scientific investigation and debate. While it presents intriguing possibilities, its confirmation would require advancements in our understanding of fundamental physics and the development of new observational techniques.

The concept of parallel universes expands our imagination and challenges our perception of reality. It opens up philosophical discussions about the nature of existence, consciousness, and our place within the cosmos. While the existence of parallel universes remains speculative at present, continued scientific exploration and theoretical developments may shed further light on this fascinating aspect of the multiverse.

- Theoretical models of parallel universes

Theoretical models of parallel universes aim to provide frameworks that explain the existence of multiple realities beyond our own. While these models are speculative and subject to ongoing scientific investigation, they offer different perspectives on how parallel universes could potentially exist. Here are some notable theoretical models:

1. Many-Worlds Interpretation (MWI): As mentioned earlier, MWI is an interpretation of quantum mechanics that suggests the existence of parallel universes. According to MWI, every quantum event results in the branching of reality, with each possible outcome being realized in a separate universe. These parallel universes exist in a superposition, each evolving independently, and encompass all possible quantum states and measurements.

2. Bubble Universes in Inflationary Cosmology: Inflationary cosmology, based on the theory of cosmic inflation, proposes that our universe underwent a period of exponential expansion shortly after the Big Bang. In this model, new regions of space can emerge and form bubble universes within a larger multiverse. Each bubble universe would have different properties and laws of physics, allowing for the existence of parallel realities.

3. Brane Worlds in String Theory: String theory, a candidate theory of quantum gravity, suggests that our universe may be a three-dimensional "brane" embedded within a higher-dimensional space. According to this model, other branes could exist alongside ours, representing separate universes or realities. These brane worlds could have different physical properties, dimensions, and fundamental forces, offering the possibility of parallel universes.

4. Mathematical Multiverse: The mathematical multiverse is a concept that arises from the idea that mathematical structures and equations describe different universes. It suggests that every mathematically consistent model corresponds to a separate universe with its own laws of physics. In this view, the existence of parallel universes is a consequence of the richness and diversity of mathematical possibilities.

5. Quantum Multiverse: The quantum multiverse hypothesis proposes that quantum superposition and entanglement give rise to parallel universes. According to this model, the wave function of the universe encompasses all possible states, with each state corresponding to a distinct parallel universe. The branching and decoherence of quantum systems result in the existence of multiple universes, each representing a different quantum state.

It's important to emphasize that these theoretical models are still speculative, and there is currently no direct observational evidence to support their existence. They are frameworks aimed at addressing fundamental questions in physics and cosmology, such as the nature of quantum mechanics, the origin of the universe, and the fundamental laws that govern reality.

Continued research, experimentation, and theoretical advancements in fields like quantum mechanics, cosmology, and string theory may provide further insights into the existence and characteristics of parallel universes.

- Evidence and speculation surrounding the existence of other realities

The existence of other realities, such as parallel universes or alternate dimensions, remains speculative, and direct empirical evidence is currently lacking. However, certain lines of scientific inquiry and theoretical considerations have motivated the exploration of these concepts. Here are some aspects related to evidence and speculation surrounding the existence of other realities:

1. Theoretical Frameworks: Theoretical frameworks like string theory, brane worlds, inflationary cosmology, and quantum mechanics provide mathematical and conceptual frameworks that allow for the possibility of other realities. While these theories are still under development and require further empirical validation, they offer avenues for exploring the existence of parallel universes or alternate dimensions.

2. Cosmological Observations: Certain cosmological observations and measurements have raised intriguing possibilities. For instance, anomalies in the cosmic microwave background radiation or patterns in large-scale structures of the universe have been interpreted as potential evidence of collisions or interactions between our universe and other universes within a multiverse scenario. However, these interpretations are speculative and subject to alternative explanations.

3. Quantum Experiments and Observations: Quantum mechanics, with its inherent uncertainty and superposition, has sparked speculation about the existence of other realities. The double-slit experiment, quantum entanglement, and delayed-choice experiments have raised questions about the nature of reality and the possibility of multiple outcomes coexisting in parallel universes or branches of reality. However, the interpretation of these quantum phenomena remains a subject of debate among physicists.

4. Mathematical Consistency: The mathematical consistency of certain theories, such as string theory or certain multiverse proposals, has been used as an argument for the existence of other realities. The internal coherence and elegance of these mathematical frameworks suggest the possibility of parallel universes or additional dimensions, even if direct empirical verification is challenging.

5. Philosophical and Conceptual Arguments: Philosophical and conceptual arguments are often employed to speculate about the existence of other realities. These arguments explore topics such as the nature of consciousness, the nature of reality beyond human perception, or the limits of our understanding. While these arguments may be thought-provoking, they do not provide direct empirical evidence.

It is essential to approach the topic of other realities with caution. While scientific theories and philosophical discussions offer intriguing possibilities, their speculative nature and the lack of direct empirical evidence mean that we are still in the realm of exploration and hypothesis. The scientific community continues to investigate and debate these concepts, aiming to develop testable predictions and gather evidence that could support or challenge the existence of other realities.

Ultimately, our understanding of other realities may evolve as scientific advancements are made, and new tools and observations become available. The pursuit of evidence and the exploration of speculation are integral to scientific progress, but for now, the existence of other realities remains an open question.

Chapter 6

Beyond the Physical: Mystical, Philosophical, and Metaphysical Perspectives

Beyond the physical realm, there are mystical, philosophical, and metaphysical perspectives that offer alternative ways of understanding reality and the existence of other realms or dimensions. These perspectives explore concepts that go beyond empirical evidence and scientific inquiry, delving into subjective experiences, consciousness, and the nature of existence. Here are some key aspects of these perspectives:

1. Mystical Experiences: Mystical traditions and spiritual practices across cultures have long suggested the existence of higher or transcendent realms beyond the physical. Mystical experiences, such as states of deep meditation, altered states of consciousness, or encounters with the divine, are believed to provide glimpses into these non-physical dimensions. These experiences are subjective and often associated with a sense of unity, interconnectedness, or expanded awareness that transcends ordinary reality.

2. Consciousness and Subjectivity: Some philosophical and metaphysical perspectives propose that consciousness plays a fundamental role in shaping reality and that it exists beyond the physical realm. These perspectives suggest that consciousness is not merely an emergent property of the brain but rather a fundamental aspect of the universe. They propose that other realities or dimensions may be accessible through shifts in consciousness or altered states of awareness.

3. Transcendentalism and Idealism: Philosophical movements like transcendentalism and idealism emphasize the primacy of the mind or consciousness in constructing reality. They propose that the physical world is an expression or manifestation of deeper spiritual or mental realities. According to these perspectives, other realms or dimensions may exist beyond the physical, and our understanding of reality is influenced by our subjective experiences and interpretations.

4. Metaphysical and Esoteric Systems: Various metaphysical and esoteric systems of thought offer frameworks for understanding reality that incorporate non-physical dimensions. These systems, such as Theosophy, Hermeticism, or Kabbalah, propose intricate cosmologies that involve hierarchies of planes, realms, or dimensions beyond the physical. These dimensions are often described as existing on subtler or higher vibrational levels and may be accessible through spiritual practices or esoteric knowledge.

5. Thought Experiments and Speculation: Philosophical and metaphysical thought experiments explore conceptual possibilities beyond empirical verification. These exercises of the imagination raise questions about the nature of reality, the existence of other realms or dimensions, and the limits of human understanding. While these thought experiments do not provide empirical evidence, they encourage contemplation and open up new avenues for exploration.

It's important to note that mystical, philosophical, and metaphysical perspectives offer alternative ways of interpreting and understanding reality, but they often operate outside the realm of scientific verification. These perspectives can be deeply personal and subjective, and interpretations may vary among individuals. They provide different lenses through which to explore the mysteries of existence and the possibility of other dimensions or realms beyond the physical.

Ultimately, the exploration of these perspectives invites reflection, introspection, and a broader consideration of the nature of reality beyond what can be observed and measured. They contribute to the rich tapestry of human thought, offering alternative perspectives and frameworks that complement scientific inquiry in our ongoing quest to understand the nature of existence.

- Timelessness in spiritual traditions and mystical experiences

In many spiritual traditions and mystical experiences, the concept of timelessness holds great significance. These traditions suggest that there is a dimension of reality beyond the linear progression of time, where a sense of timelessness is experienced. Here are some key aspects related to timelessness in spiritual traditions and mystical experiences:

1. Eternal Present: Spiritual traditions often emphasize the importance of being fully present in the moment, transcending the limitations of past and future. The eternal present is seen as a state of consciousness where one becomes aware of the timeless nature of reality. This can be achieved through practices such as meditation, mindfulness, or deep contemplation.

2. Transcending Time: Mystical experiences often involve a sense of timelessness, where ordinary concepts of past, present, and future fade away. During these experiences, individuals may report a merging with a greater reality or a sense of oneness with the universe, transcending the constraints of time. These experiences are described as a profound connection with a timeless and infinite aspect of existence.

3. Non-Dual Awareness: Many spiritual traditions emphasize the realization of non-dual awareness, where the boundaries between self and the external world dissolve. In this state, the sense of separate individuality and the linear flow of time are transcended, leading to a direct experience of the timeless nature of consciousness or existence itself.

4. Divine Timelessness: Some spiritual traditions suggest that the divine or ultimate reality exists beyond the realm of time. It is considered timeless and eternal, not subject to the limitations and constraints of temporal existence. From this perspective, the divine is seen as an unchanging presence that encompasses all of creation, past, present, and future.

5. Symbolic Timelessness: Symbolic narratives and myths in various spiritual traditions often convey timeless truths or archetypal themes that resonate across cultures and generations. These stories, rituals, and symbols are believed to tap into timeless aspects of the human experience and offer insights into the nature of reality beyond the constraints of ordinary time.

It is important to note that the experience of timelessness in spiritual traditions and mystical experiences is subjective and personal. It often defies easy explanation or empirical verification. However, for those who have undergone such experiences, the sense of timelessness can be deeply transformative, providing a profound shift in perspective and a connection to a greater reality beyond the temporal.

Exploring the concept of timelessness in spiritual traditions and mystical experiences invites us to consider the nature of time itself and our relationship with it. It challenges our conventional understanding of time as a linear progression and invites us to explore the depths of present moment awareness and the timeless dimensions of existence.

- Philosophical ponderings on the nature of time and its implications

The nature of time has been a subject of philosophical pondering for centuries, leading to a variety of perspectives and implications. Philosophers have contemplated questions such as the nature of time's flow, its relationship to causality, its existence independent of human perception, and its implications for free will and personal identity. Here are some key philosophical ponderings on the nature of time and its implications:

1. The Nature of Time: Philosophers have debated whether time is an objective feature of the external world or merely a subjective construct of human consciousness. Some argue that time exists independently of human perception, while others propose that it is a product of our cognitive processes and relational experiences.

2. The Arrow of Time: The concept of the "arrow of time" explores the directionality of time's flow. Philosophers contemplate whether time has a fundamental direction, typically associated with the asymmetry between past and future. The arrow of time raises questions about the irreversibility of events, the nature of causality, and the possibility of time travel.

3. Eternalism vs. Presentism: Philosophers debate two major positions regarding the ontology of time: eternalism and presentism.

Eternalism posits that all moments in time are equally real and that past, present, and future exist simultaneously. Presentism, on the other hand, suggests that only the present moment is real, and past and future are mere abstractions or potentialities.

4. Free Will and Determinism: The nature of time has implications for the philosophical debate surrounding free will and determinism. If time is viewed as a fixed and predetermined sequence of events, it raises questions about the extent of human agency and the possibility of true freedom. Philosophers explore whether human actions are causally determined or if there is room for genuine choice and autonomy within the framework of time.

5. Personal Identity and Temporal Continuity: Philosophers examine how our experience of time shapes our sense of personal identity. Questions arise regarding the persistence of personal identity over time and whether it relies on a continuous temporal stream or some other factors, such as memories, psychological continuity, or narratives of self.

6. Time and Experience: Philosophers explore the relationship between time and our subjective experience of reality. They ponder the nature of temporal perception, the role of memory in shaping our understanding of the past, and how our anticipation of the future influences our present actions and decisions.

7. Timelessness and the Transcendent: Some philosophical perspectives consider the possibility of timeless or eternal dimensions beyond the ordinary flow of time. These perspectives examine concepts of transcendence, divine existence, and the timeless nature of ultimate reality.

These philosophical ponderings on the nature of time are complex and have implications that reach into various areas of philosophy, such as metaphysics, epistemology, ethics, and philosophy of mind. They encourage deep reflection on our understanding of reality, the nature of existence, and our place within the fabric of time.

It's important to note that these philosophical inquiries often generate more questions than definite answers. The exploration of the nature of time continues to evolve, incorporating insights from scientific discoveries, philosophical debates, and interdisciplinary perspectives.

Chapter 7

The Future of Time and Space: Speculations and Discoveries

The future of time and space is a fascinating subject that encompasses both speculative ideas and ongoing scientific discoveries. While it is impossible to predict with certainty what the future holds, here are some speculations and areas of exploration that could shape our understanding of time and space:

1. Quantum Gravity: One of the biggest challenges in theoretical physics is the development of a consistent theory that combines quantum mechanics and general relativity. This theory, often referred to as quantum gravity, could revolutionize our understanding of space, time, and the fundamental nature of reality. Various approaches, such as string theory, loop quantum gravity, and others, are being pursued in the quest to unravel the mysteries of quantum gravity.

2. Multiverse and Parallel Realities: The concept of a multiverse, which posits the existence of multiple universes or parallel realities, continues to be explored. Theoretical models, such as the inflationary multiverse or the many-worlds interpretation of quantum mechanics, suggest the possibility of other universes or dimensions beyond our own. Ongoing research, cosmological observations, and advances in theoretical physics may provide further insights into the existence or nature of these parallel realities.

3. Spacetime Engineering: Speculative ideas about manipulating spacetime have been proposed in science fiction and theoretical physics. Concepts like wormholes, warp drives, or the manipulation of spacetime curvature raise questions about the potential for future technologies to enable traversable paths through space and time. While these ideas remain highly speculative, they inspire scientific investigations into the fundamental properties of spacetime and the possibilities of manipulating them.

4. Temporal Paradoxes and Resolution: Time travel and its associated paradoxes, such as the grandfather paradox or the bootstrap paradox, continue to intrigue both scientists and philosophers. The resolution of these paradoxes, if time travel were ever to become a reality, remains an open question. Exploring the implications of time travel may lead to deeper insights into the nature of causality, the structure of time, and the boundaries of possibility.

5. Cosmological Discoveries: Advancements in observational cosmology, such as the study of the cosmic microwave background radiation, the discovery of new celestial objects, or the mapping of the large-scale structure of the universe, could provide further clues about the nature of space and time. Future missions, telescopes, and instruments may unveil new cosmic phenomena that challenge our current understanding and open up avenues for deeper exploration.

6. Philosophical and Cultural Perspectives: The future of time and space is not solely dependent on scientific advancements. Philosophical and cultural perspectives will continue to shape our understanding and exploration of these concepts. Evolving philosophical debates, cultural shifts, and interdisciplinary collaborations can offer fresh insights and perspectives on the nature of time, space, and our relationship to them.

As scientific knowledge expands and technological advancements continue, our understanding of time and space will likely evolve. The interplay between scientific discoveries, theoretical speculations, philosophical contemplations, and cultural perspectives will shape the future of our exploration into the mysteries of time and space. It is an ongoing journey of discovery and wonder that will continue to captivate the imagination of both scientists and thinkers across various disciplines.

- Cutting-edge research and breakthroughs in understanding time

Cutting-edge research in understanding time encompasses various scientific disciplines and theoretical frameworks. While our knowledge is continually evolving, here are some notable areas of research and recent breakthroughs:

1. Quantum Time: Researchers are exploring the nature of time within the framework of quantum mechanics. Quantum theories of time aim to reconcile the inherent indeterminacy of quantum physics with the seemingly continuous and deterministic nature of time. Recent studies have investigated quantum clocks, time dilation in quantum systems, and the relationship between quantum entanglement and time.

2. Time Perception and the Brain:
Neuroscientists are investigating how the brain processes and perceives time. Studies using neuroimaging techniques have shed light on the neural mechanisms underlying our sense of time and temporal processing. Research has also explored the connection between time perception and consciousness, exploring how our subjective experience of time is constructed by the brain.

3. Cosmological Time: Advancements in observational cosmology, such as the measurement of the cosmic microwave background radiation and the study of the large-scale structure of the universe, have provided insights into the nature of cosmic time. Research in cosmology aims to understand the origin and evolution of the universe, including the early stages of cosmic inflation and the formation of structures over billions of years.

4. Time Crystals: Time crystals are a recently discovered phase of matter that exhibits temporal order and symmetry breaking. These unique structures oscillate between different states in time without the need for external energy. The study of time crystals is a burgeoning field that may lead to new insights into the fundamental nature of time and its connection to the behavior of matter.

5. Temporal Causality and Retrocausality: Researchers are exploring the nature of causal relationships in the context of time. Some theories and experiments have challenged the notion of causality flowing strictly from past to future and have suggested the possibility of retrocausality, where future events can influence past events. Investigations in this area aim to deepen our understanding of the fundamental principles underlying cause and effect.

6. Timekeeping Technologies: Advancements in timekeeping technologies, such as atomic clocks and optical frequency combs, have allowed for highly accurate measurements of time and frequency. These technological breakthroughs are crucial for various fields, including GPS navigation, telecommunications, and fundamental physics research. They provide precise tools for studying the nature of time and its applications in practical domains.

It is important to note that the study of time is interdisciplinary, involving fields such as physics, neuroscience, philosophy, and mathematics. Breakthroughs in one field often have implications for others, leading to new insights and collaborative efforts. The ongoing research in these areas and the interplay between different disciplines continue to shape our understanding of time, bringing us closer to unraveling its mysteries.

- Exciting possibilities and unanswered questions on the horizon

As we delve deeper into the study of time, numerous exciting possibilities and unanswered questions emerge, spurring further exploration and discovery. Here are some of the captivating possibilities and open questions on the horizon:

1. Quantum Gravity and Spacetime: The quest for a theory of quantum gravity remains a prominent challenge in theoretical physics. Discovering a unified framework that reconciles quantum mechanics and general relativity could unveil a deeper understanding of the nature of spacetime, gravity, and the fabric of the universe.

2. Time Travel: While time travel remains speculative, ongoing research into the physics and theoretical frameworks opens up intriguing possibilities. Exploring the feasibility, paradoxes, and implications of time travel could shed light on fundamental aspects of time and causality, pushing the boundaries of our understanding.

3. Nature of Temporal Flow: The question of why time appears to flow in a particular direction, from past to future, is still a puzzle. Investigating the arrow of time and understanding its origins, especially in relation to the microscopic laws of physics, presents an intriguing avenue for future exploration.

4. Consciousness and Time: The relationship between subjective experience, consciousness, and time is an area ripe for further investigation. Understanding how our conscious perception shapes our experience of time, and vice versa, could provide insights into the nature of consciousness and the foundations of our temporal reality.

5. Temporal Boundaries and Singularities: Exploring the nature of singularities, such as those found in black holes or at the beginning of the universe, presents captivating possibilities. Investigating what lies beyond these boundaries and how they impact our understanding of time could unlock new frontiers in cosmology and theoretical physics.

6. Interplay of Time and Information: The connection between time and information processing is an intriguing field of research. Examining how information is stored, processed, and transformed over time, particularly in biological systems and quantum computing, could lead to breakthroughs in understanding the relationship between information, entropy, and the arrow of time.

7. Time and Conscious AI: As artificial intelligence progresses, questions about the relationship between time and consciousness in AI systems arise. Investigating how AI perceives and experiences time, and whether consciousness can be replicated or simulated, opens up philosophical and ethical inquiries into the nature of sentience and the boundaries of machine intelligence.

These exciting possibilities and unanswered questions demonstrate the vastness of the field of time studies and its interdisciplinary nature. The ongoing collaboration between physicists, neuroscientists, philosophers, and other scholars offers promising avenues for future exploration, pushing the boundaries of our knowledge and unraveling the mysteries of time.

Conclusion

Embracing the Enigma

Embracing the enigma of time and space is an invitation to appreciate the profound mysteries that surround us. Despite centuries of scientific and philosophical exploration, many aspects of time and space remain elusive and awe-inspiring. Embracing the enigma means acknowledging that there are limits to our current understanding and being open to the possibility of new discoveries that may challenge our existing concepts.

The enigma of time and space invites us to embrace curiosity, wonder, and a sense of humility in the face of the unknown. It encourages us to ask bold questions, explore different perspectives, and engage in ongoing intellectual and scientific inquiry. Rather than feeling discouraged by the enigma, we can find inspiration in its vastness and the limitless possibilities it presents.

Embracing the enigma also reminds us of the interconnectedness of knowledge. The study of time and space transcends disciplinary boundaries, requiring collaboration and integration of insights from physics, philosophy, neuroscience, cosmology, and more. By embracing the enigma, we can foster interdisciplinary dialogue and appreciate the richness that arises from diverse perspectives.

Furthermore, the enigma of time and space holds a sense of wonder and mystery that can inspire artistic expression, spiritual contemplation, and personal introspection. It invites us to reflect on our place in the cosmos, our fleeting existence within the grand tapestry of time, and the significance of our experiences.

Ultimately, embracing the enigma of time and space is an acknowledgment that there will always be more to explore, more questions to ask, and more mysteries to unravel. It is an invitation to embark on a continuous journey of discovery, both as individuals and as a collective, and to revel in the profound beauty and complexity of the universe in which we find ourselves.

- Reflecting on the journey and the wonders uncovered

Reflecting on the journey of exploring time and space fills us with a sense of awe and wonder at the wonders we have uncovered thus far. It is a testament to the incredible capacity of human curiosity and intellect to unravel the mysteries of the universe. Here are some reflections on the journey and the wonders uncovered:

1. Expanding Horizons: Our understanding of time and space has expanded exponentially throughout history. From ancient civilizations contemplating the cyclical nature of time to groundbreaking scientific theories like relativity and quantum mechanics, each milestone has pushed the boundaries of our knowledge and opened new realms of exploration.

2. Deep Connections: The exploration of time and space has revealed the profound interconnections between different fields of study. Physics, philosophy, neuroscience, and cosmology, among others, have converged to shed light on the nature of existence, consciousness, and our place in the vast cosmic web. It highlights the beauty of interdisciplinary collaboration and the power of shared knowledge.

3. Astonishing Discoveries: Along the journey, we have encountered astonishing discoveries that have reshaped our understanding of reality. From the mind-bending implications of relativity and quantum entanglement to the mind-expanding concepts of parallel universes and time dilation, each revelation has challenged our preconceptions and expanded the realm of what is possible.

4. Humbling Perspectives: The exploration of time and space humbles us in the face of the immense complexity and grandeur of the universe. It reminds us of our limited perspective and the vastness of the unknown that lies ahead. It is a humbling reminder that the more we discover, the more we realize how much more there is to learn.

5. Inspiring Wonder: Reflecting on the wonders uncovered evokes a sense of wonder and curiosity, fueling our desire to keep exploring. It ignites our imagination and inspires us to continue pushing the boundaries of knowledge, driven by the innate human thirst for understanding and discovery.

6. Philosophical and Existential Insights: The journey into time and space has also yielded profound philosophical and existential insights. It prompts contemplation on the nature of existence, the mystery of consciousness, and the fleeting nature of our own lives. It invites us to ponder the profound questions about our purpose, our interconnectedness, and the meaning we assign to our experiences.

7. Endless Exploration: The journey into time and space is far from complete. There are still countless mysteries awaiting our investigation. The reflection on the wonders uncovered reminds us that there is always more to explore, discover, and understand. It encourages us to remain curious, open-minded, and committed to the ongoing pursuit of knowledge.

In reflecting on the journey and the wonders uncovered, we are reminded of the beauty and complexity of the universe we inhabit. It instills in us a sense of gratitude for the remarkable insights gained and the excitement of the unknown that lies ahead. It is a celebration of the human spirit's relentless quest for understanding and the profound impact it has had on shaping our perception of the world.

- Embracing the mysteries that continue to captivate our imaginations

Embracing the mysteries that continue to captivate our imaginations is an invitation to acknowledge and appreciate the inherent beauty and intrigue of the unknown. Despite the remarkable progress we have made in understanding time and space, there are still enigmatic aspects that spark our curiosity and inspire us to delve deeper. Here are some thoughts on embracing these captivating mysteries:

1. Fueling Curiosity: The mysteries that persist in the realm of time and space fuel our curiosity, driving us to explore and seek answers. They encourage us to ask thought-provoking questions and push the boundaries of our knowledge. Embracing these mysteries nurtures our sense of wonder and fosters a lifelong thirst for learning and discovery.

2. Stimulating Imagination: Unanswered questions and enigmatic phenomena stimulate our imagination, inviting us to envision new possibilities and scenarios. They serve as fertile ground for scientific hypotheses, creative thinking, and speculative ideas. Embracing these mysteries encourages us to think beyond the confines of our current understanding and envision alternative explanations and realities.

3. Inspiring Scientific Advancement: The mysteries that captivate our imaginations often serve as driving forces behind scientific advancement. They motivate researchers to develop innovative theories, design experiments, and devise new methods of investigation. By embracing these mysteries, we create fertile ground for scientific progress and the pursuit of knowledge.

4. Cultivating Humility: Embracing the mysteries reminds us of the vastness and complexity of the universe. It humbles us, reminding us that there are limits to our understanding and that there is always more to learn. It fosters a sense of humility, encouraging us to approach the mysteries with an open mind and a willingness to explore different perspectives.

5. Encouraging Collaboration: Embracing the mysteries of time and space invites collaboration and the sharing of ideas across disciplines. These mysteries often transcend the boundaries of individual fields of study, requiring interdisciplinary approaches to unravel. By embracing these mysteries collectively, we can pool our expertise and perspectives, leading to deeper insights and breakthroughs.

6. Nurturing a Sense of Awe: The mysteries that captivate our imaginations evoke a sense of awe and wonder. They remind us of the inherent beauty and intricacy of the universe, instilling a sense of reverence for the mysteries that remain unsolved. Embracing these mysteries nurtures our appreciation for the grandeur of the cosmos and our place within it.

By embracing the mysteries that continue to captivate our imaginations, we embark on a never-ending journey of exploration and discovery. It is an invitation to approach the unknown with curiosity, imagination, and humility, acknowledging that the pursuit of knowledge is an ongoing and transformative process. Embracing these mysteries enriches our lives, deepens our understanding of the world, and fuels our innate sense of wonder.

www.ingramcontent.com/pod-product-compliance
Lightning Source LLC
Chambersburg PA
CBHW071129240526
45465CB00024B/1548